YOU
CAN SAVE THE
EARTH

A Handbook for Environmental
Awareness, Conservation,
and Sustainability

Hatherleigh Press is committed to preserving and protecting the natural resources of the earth. Environmentally responsible and sustainable practices are embraced within the company's mission statement.

Visit us at www.hatherleighpress.com and register online for free offers, discounts, special events, and more.

You Can Save the Earth, Revised Edition
Text Copyright © 2018 Hatherleigh Press

Library of Congress Cataloging-in-Publication Data is available upon request.
ISBN: 978-1-57826-670-8

A Prayer for the World © 2008 Amy Powers, Rob Wells and Charlene Gilliam

Hatherleigh Press is committed to preserving and protecting the natural resources of the Earth. Environmentally responsible and sustainable practices are embraced within the company's mission statement.

Hatherleigh Press is a member of the Publishers Earth Alliance, committed to preserving and protecting the natural resources of the planet while developing a sustainable business model for the book publishing industry.

Cover and interior design by Carolyn Kasper

Printed in the United States
10 9 8 7 6 5 4 3 2 1

Mankind has long endeavored to reach out to the heavens. We have set foot on the moon, launched probes into the vast depths of space, pondered dwellings in zero gravity. Yet, perhaps all this effort and investment was spent not to find a new home, but rather to turn the cameras around and capture a photo of our Earth. Beautiful, fragile, alive. A solitary Ark carrying us forward through time.

We are at a watershed moment in the history of our world, when we must make the changes required to embrace a sustainable future. The good news is that it's a choice we are free to make, and we should do so with a spirit of optimism.

Consider *You Can Save the Earth* a gift of hope and an enduring message to help safeguard the Earth, its environments, and each other.

—Andrew Flach, Publisher

CONTENTS

PREFACE TO THE REVISED EDITION

Ten years have passed since the publication of the first edition of *You Can Save the Earth*. Much has changed on the environmental front, for better and (unfortunately) for worse.

"Sustainability" has become a buzzword and a public policy goal. Alternative fuel and fuel-efficient vehicles are now viable, affordable options for most consumers. Water-conservation initiatives (including device distribution, and rebate and voucher programs) are in force in municipalities nationwide. Over 150 American cities or counties have instituted plastic bag bans or fees. And farmers markets featuring locally-sourced fruits, vegetables, and other foodstuffs are springing up everywhere; as a rebuttal to the "food miles" and "factory farm" debates, farmers markets not only are truly local endeavors (more than half of farmers traveled less than 10 miles to their market), but they are directly responsible for the renewed growth of family farms and local agricultural economies.

Despite these obvious achievements, however, our earth has not become a healthier habitat overall. Global warming continues unabated, causing climate change and its many deleterious offshoots, including higher temperatures, drought, rising seas, melting glaciers, and famine. Vector-borne diseases thrive in regions they might have once found inhospitable—just this summer, mosquitoes carrying the Zika virus have been discovered in southern Florida. Fracking (hydraulic fracturing) pollutes groundwater reservoirs and produces dangerous amounts of methane. Elsewhere, municipal water systems have been compromised by agricultural and industrial waste, as well as an aging storage and transportation infrastructure; systemic failures (like Flint, Michigan) are sure to become more common.

Ten years later, our planet is still under threat . . . a mortal threat. Which makes this book's core message—of caring and consciousness, community and civic involvement—more relevant than ever.

"If future generations are to remember us with gratitude rather than contempt, we must leave them something more than the miracles of technology," observed President Lyndon Johnson in 1964 as he signed the Wilderness Act into law. "We must leave them with a glimpse of the world as it was in the beginning, not just after we got through with it."

As neighbors, as citizens, and most importantly, as human beings, we must roll up our sleeves and set about the all-important task of saving our one and only home.

"It seems to me that the natural world is the greatest source of excitement; the greatest source of visual beauty; the greatest source of intellectual interest. It is the greatest source of so much in life that makes life worth living."

—DAVID ATTENBOROUGH

YOU

CAN SAVE THE EARTH

This book is about how you can save the Earth.

The threat to Earth's environment isn't new. But it is more pressing than it has ever been. There is no doubt that our way of living is causing the Earth tremendous damage. This is damage we must stop—for ourselves, and for the generations that will come after us.

However, this book does not contain photographs of polluted skies or trees stripped bare by acid rain or of overflowing landfills. Certainly, you have seen enough of those images.

"Conservation is a state of harmony between men and land . . . harmony with land is like harmony with a friend; you cannot cherish his right hand and chop off his left."

—ALDO LEOPOLD

And if those photos haven't already motivated you, then they never will.

Instead of terrifying you, this book is going to empower you.

Begin by understanding how our lifestyle is affecting the Earth we inhabit. Then, think about why *you* value clear skies, safe water, and a bountiful natural landscape. Suddenly, your own motivation will become clear to you. Once you are ready to act, learn about steps you can take that will truly make a difference for the Earth, and for all of us.

You can save the Earth.

"The care of the Earth is our most ancient and most worthy . . . responsibility. To cherish what remains of it and to foster its renewal is our only hope."

—WENDELL BERRY

7 REASONS WHY

"The love of wilderness is more than a hunger for what is always beyond reach; it is also an expression of loyalty to the earth, the earth which bore us and sustains us, the only paradise we shall ever know, the only paradise we ever need . . . wilderness is not a luxury but a necessity of the human spirit, as vital to our lives as water and good bread."

—EDWARD ABBEY

YOU

CAN SAVE THE EARTH

How is the way we live now affecting the Earth? It's important to understand why our means of transportation, the way we eat, and the energy we consume are changing our environment, so we can protect our planet—our one and only home.

As you learn more about the effect of our habits on the Earth, think about why a healthy environment is important to you. Remind yourself how much you value clean air, clear water, litter-free beaches, fresh food and unspoiled open spaces. Aren't these part of what makes life worth living?

"All that we did, all that we said or sang must come from contact with the soil."

—William Butler Yeats

Find your own motivation to guard a healthy Earth. Maybe it's because you love to fly fish in sparkling streams. Maybe you're fond of morning bike rides or walking your dog through the woods. Let your own personal experiences guide you.

Listen to your own voice—your own conscience—when you hear more about the state of the environment around the globe.

"If, as the elders have told us, we are our grandparents' dream, then we must today begin dreaming of our grandchildren."

—WALTER BRESETTE

1

ALL OF LIFE IS INTERCONNECTED

"When we try to pick out anything by itself, we find it hitched to everything else in the Universe."

—JOHN MUIR

ALL OF LIFE on Earth is connected.

A seed needs soil, sun, and water to grow into a vegetable. The vegetable provides our bodies with the energy needed to function and thrive.

We must guard this precious cycle of life by protecting every element at each stage of growth. For, if one component is threatened, we are all at risk.

Here is an example: In 2015, underground peat fires in Indonesia created a cloud of toxic smoke which stretched not only into neighboring Singapore and Malaysia, but halfway around the globe. It's estimated that in just two months, the Indonesian fires emitted some 1.5 billion metric tons of CO_2 into the atmosphere—more than Japan's total fossil fuel emissions for 2013.

When the quality of the air is compromised somewhere in the world, we all pay the price. The Earth does not recognize borders or nationalities. This means our neighbor's environmental disasters are also ours.

Runoff from the farming states that drain

"How we spend our days is, of course, how we spend our lives."

—ANNIE DILLARD

into the Mississippi River contains unhealthy levels of nitrogen and phosphorus (mostly from crop fertilizers and manure). This toxic soup flows down the Mississippi and into the Gulf of Mexico, where it has created the world's second largest "dead zone"—a vast aquatic desert in which dying algae depletes the water's oxygen and literally suffocates marine life. Early estimates held that the 2016 Gulf dead zone would span 6,800 square miles (roughly the size of the state of Connecticut). 43 of the world's 166 dead zones are located in the United States, among them the scenic Chesapeake Bay. Globally, dead zones can be found in the Baltic Sea, the northern Adriatic Sea, the Yellow Sea, and the Gulf of Thailand.

All of life on Earth is connected by actions across time and place. Our choices here, today, shape the world that our children and grandchildren will inherit.

Our descendants rely on us all to preserve our only home and leave it for them in better shape than we found it.

All of us must work to preserve life on Earth. Saving the Earth depends on the cooperation of communities, both locally and

"You see that pale, blue dot? That's us. Everything that has ever happened in all of human history, has happened on that pixel. All the triumphs and all the tragedies, all the wars all the famines, all the major advances . . . it's our only home. And that is what is at stake, our ability to live on planet Earth, to have a future as a civilization. I believe this is a moral issue, it is your time to seize this issue, it is our time to rise again to secure our future."

—AL GORE

globally. We are all responsible for protecting our air, water, and soil.

Recognizing our interdependence is the first step to saving our planet. We humans have only one home, and we must care for it—together.

EXERCISE:
 Find a quiet place. Close your eyes for a moment and feel yourself connected to all living things.

AFFIRMATION:
 I am connected to life and to our Earth.

"The first wealth is health."

—RALPH WALDO EMERSON

2

THE HEALTH OF THE EARTH DEFINES OUR WELL-BEING

"Today, the world population is encounting unfamiliar human-induced changes in the lower and middle atmospheres and world-wide depletion of various other natural systems. Beyond the early recognition that such changes would affect economic activities, infrastructure and managed ecosystems, there is now recognition that global climate change poses risks to human population health."

—WHO ON CLIMATE CHANGE AND HUMAN HEALTH

THE HEALTH OF the Earth is vital to our health. If the soil is poisoned, a plant will not grow. If our air is polluted, our health is at risk.

The consequences for our health from global warming and climate change are numerous. The World Health Organization reports, "Climate change can affect human health directly (e.g., impacts of thermal stress, death/injury in floods and storms) and indirectly through changes in the ranges of disease vectors (e.g., mosquitoes), water-borne pathogens, water quality, air quality, and food availability and quality."

Freezes and flooding, and droughts and heat waves, result in multiple deaths and outbreaks of disease. Physical displacement due to cata-clysmic weather events (hurricanes, tsunamis) often results in contaminated drinking water supplies which can lead to massive outbreaks of cholera and other waterborne diseases.

As the Earth's temperatures rise, there has been a simultaneous increase in outbreaks of certain heat-related infectious diseases,

"We pray that we may at all times keep our minds open to new ideas and shun dogma; that we may grow in our understanding of the nature of all living beings and our connectedness with the natural world; that we may become ever more filled with generosity of spirit and true compassion and love for all life; that we may strive to heal the hurts that we have inflicted on nature and control our greed for material things; [...] that we may value each and every human being for who he is, for who she is, reaching to the spirit that is within, knowing the power of each individual to change the world."

—JANE GOODALL

including malaria, dengue fever, and yellow fever. This occurs often in regions that had not previously seen the diseases—including the United States, where cases of the West Nile virus and Lyme disease have appeared. Air pollution and declining air quality have resulted in a doubling if not a tripling of asthma cases over the past 20 years. Changing weather patterns affect agriculture; threatened crops and livestock can result in widespread undernourishment. According to a recent World Health Organization report, 2015 "will go down in history as the hottest year on record, and such extreme weather can be deadly; droughts, storm surges, heatwaves, wildfires and floods claim thousands of human lives and livelihoods. WHO estimates that more than 7 million deaths worldwide can be attributed to air pollution alone every year, and that by 2030, climate change will be causing hundreds of thousands of additional deaths from malaria, [diarrheal] disease, heat stress and under nutrition. Climate change also plays a clear role in the emergence of new diseases and outbreaks, such as the Zika virus." As WHO Director-General Dr. Margaret Chan warns,

"The future is knocking at our door right now. Make no mistake, the next generation will ask us one of two questions. Either they will ask, 'What were you thinking; why didn't you act?' or they will ask instead, 'How did you find the moral courage to rise and successfully resolve a crisis that so many said was impossible to solve?'"

—AL GORE

"A ruined planet cannot sustain human lives in good health.

Once, miners used caged canaries to monitor the levels of odorless poisonous gases in mine shafts. If the caged canary was singing and strong, the shaft's air was safe to breathe. But if the canary's song stopped, the miners knew the air was deadly and that they were in grave danger.

Those who are ill because of air pollution and declining air quality are showing us that our health is threatened. This is a warning to us that we must reverse course—that we must reconsider our actions to protect the health of the Earth.

For us to be healthy, our Earth must be healthy. The time has come to protect our health.

EXERCISE:
Gently place your fingers on your pulse and acknowledge the extraordinary miracle of your beating heart.

AFFIRMATION:
I value my health. I value my family's health. I value the Earth's health.

"We forget that the water cycle and the life cycle are one."

—JACQUES COUSTEAU

3

WATER IS THE ESSENCE OF ALL LIFE

"Rivers are roads which move,
and which carry us whither
we desire to go."

—BLAISE PASCAL

WE OFTEN FORGET that our first stages of life unfold "underwater," in the womb.

After we enter the world, water continues to enable our growth. We need water in order to live.

Yet, the *way* we live is threatening this vital resource. Here is an example: bottled water. According to some estimates, Americans consumed 11 billion gallons of bottled water in 2014—that's the equivalent of 34 gallons per person. Although we are doing our body good by consuming our daily intake of H_2O, purchasing water in plastic bottles is a big mistake for our Earth. In 2015, we used about 50 billion plastic water bottles. But our recycling rate for plastic is only about 23 percent—which means some 38 billion water bottles never reach the recycling center, but are sent to our already overcrowded landfills or end up as trash.

Second, manufacturing those plastic bottles requires the use of an expensive,

"A river seems a magic thing.
A magic, moving, living part
of the very earth itself."

—Laura Gilpin

nonrenewable, and polluting resource: oil. It's estimated that it takes 17 million barrels of oil to produce those 50 billion plastic water bottles—that's enough oil to fuel more than 1 million cars and trucks *for a whole year.* And all that oil is just for manufacturing the bottles; even more is required for shipment, delivery, and removal. One estimate holds that it takes a quarter bottle of oil to produce and deliver every bottle of water.

Oil leads to pollution, and the effects of pollution *aren't* just limited to some faraway shore. Pollution has come home, into the water and onto the shorelines we love to enjoy. Nearly half of America's rivers and lakes are too polluted for fishing or swimming. That's because our recreational waters are polluted by over a trillion gallons of untreated sewage, storm water, and industrial waste every year. This also threatens our drinking water supplies, since sewage and industrial waste can leach into our groundwater and wells.

Our waters are also poisoned by air pollution. Greenhouse emissions in the atmosphere (specifically excess carbon dioxide) can bind with seawater, leading to the dangerous acidification of our oceans, which is devastating

"Man is not an aquatic animal,
but from the time we stand
in youthful wonder beside a
spring brook till we sit in old
age and watch the endless roll
of the sea, we feel a strong
kinship with the waters of this
world."

—HAL BORLAND

our coral reefs and endangering the aquatic food web.

If you take a moment to think about some of the most popular vacation spots around the world, you will probably think of several getaways near a body of water. Oceans, lakes, streams, and waterfalls relax our bodies and cleanse our minds. Whether we seek water to swim, gaze at the waves, or listen to the rhythm of the surf, water relieves stress and elevates our state of mind.

Water is a source of healing. We must use this precious element responsibly and guard its cleanliness for our future. It is important not only for the health of our bodies, but for the health of our sprits as well.

We must ensure the future of our clean drinking water, and we must preserve the bodies of water that we so love to admire.

"The best thing one can do when it's raining is to let it rain."

—HENRY WADSWORTH
LONGFELLOW

EXERCISE:

Pour yourself a glass of water in a clear container. For a moment, reflect on how your own body is made up of over 50% percent water.

AFFIRMATION:

Water flows within me. I cherish pure water.

"I only went out for a walk and finally concluded to stay out till sundown, for going out, I found, was really going in."

—John Muir

4

THE AIR WE
BREATHE

"The time has come to lower our voices, to cease imposing our mechanistic patterns on the biological processes of the earth, to resist the impulse to control, to command, to force, to oppress, and to begin quite humbly to follow the guidance of the larger community on which all life depends."

—THOMAS BERRY

HOW MANY BREATHS do we take in a day? It would be impossible to count. We cannot live without breathing. Air is our lifeline.

For this reason, we must stop polluting our skies.

Statistics suggest that the United States is responsible for 16% of the world's greenhouse gas emissions, a result of our overwhelming dependence on fossil fuels.

We burn fossil fuels to run our cars and planes (gasoline and oil) and light our homes (electricity from coal and other sources). But it's our centuries-long reliance on oil that is the major culprit here. Oil accounts for 92% of fuel consumption in the U.S. transportation sector and 40% in the industrial sector. When we burn fossil fuels to produce energy, we generate massive quantities of carbon dioxide and other gases like nitrous oxide, ozone, and methane. Automobile exhaust is 95 percent carbon dioxide. In 2014, cars, motorcycles, trucks, and buses drove nearly 3 trillion miles in our country—farther than driving to the

"This most excellent canopy, the air . . . this brave o'erhanging firmament, this majestical roof . . ."

—WILLIAM SHAKESPEARE

sun and back 16,000 times. Since the average automobile releases more than 19 pounds of CO_2 and other greenhouse gases per gallon of gasoline, this means that each of our personal vehicles is producing over 5 metric tons of carbon dioxide per year. These numbers add up even faster when you look at the energy we also burn for planes, trains, and trucks: almost 6 million metric tons of carbon dioxide annually.

Among other things, heavy pollution in the air causes: human heart and lung disease, including asthma and emphysema, damage to crop plants and trees, sometimes in the form of acid rain, structural damage to some of our national monuments are decaying at rapid rates because of smog from cities and water pollution, when mercury and other heavy metals are absorbed by rivers and ponds. Research indicates that half of all Americans live in areas with unsafe levels of air pollution.

Every one of us needs to be more aware of how much pollution we add to the air. Think about it: when we can no longer see the "purple mountain majesties" of our national anthem, then it's time for a change.

"The Earth is what we all have in common. It is what we are made of and what we live from, and we cannot damage it without damaging those with whom we share it."

—WENDELL BERRY

EXERCISE:
Wherever you are right now, breathe deeply and fully, filling you lungs with air. Exhale slowly. Be conscious of each breath.

AFFIRMATION:
Though invisible, the air I breathe sustains my life. I value fresh, clean air.

"The United States has the world's mightiest economy and most mobile society. Yet the oil that fueled its strength has become its greatest weakness."

—AMORY LOVINS

5

THE LIMITS OF
FOSSIL FUELS

"Oil is seldom found where it is most needed, and seldom most needed where it is found."

—L.E.J. Brouwer

OIL HAS BEEN an essential tool for nearly everything our society has accomplished. Our country's history and that of much of the world, has been shaped by oil.

Cheap and abundant oil and coal have made our prosperity possible. It was the discovery of coal deposits and the technology to use that energy that made the Industrial Revolution possible. Then, the oil boom led to an age of invention. But after 200 years of nonstop oil and coal use, the Earth's limited supply of readily-accessible fossil fuels is running low. Prominent energy expert Richard Heinberg sums it up perfectly: "We are today living at the end of the period of greatest material abundance in human history—an abundance based on temporary sources of cheap energy that made all else possible." Now we must invent a new, better way to live.

We must also stop the damage to our land caused by oil drilling.

In our relentless search for untapped oil reserves, we drill in wilderness areas,

"You cannot change the fruits
that are already hanging on
the tree. You can, however,
change tomorrow's fruits.
But to do so, you will have
to dig below the ground and
strengthen the roots."

—T. HARV EKER

threatening the integrity of these unspoiled corners of our planet. Oil tanker accidents spill millions of gallons into the ocean, damaging miles of beaches and killing or injuring so many of the birds, fish and sea mammals that we love.

"Fracking" (or hydraulic fracturing), now allows us to search deep beneath the earth's surface for otherwise hidden pockets of natural gas and oil. The dictionary defines fracking as "a technique in which a liquid is injected under high pressure into a well in order to create tiny fissures in the rock deep beneath the earth which then allow gas and oil to flow into the well." Usually, that "liquid" is a slurry of local ground water mixed with highly toxic chemicals. In the United States, fracking has permitted the exploitation of previously inaccessible stores of domestic gas and oil. But this comes with a very high price tag in terms of health risks and environmental damage. According to EPA estimates, between 2000 and 2013, more than 9 million people lived within a mile of a hydraulically fractured well, with about 6,800 drinking water supplies located within 1 mile of a fracking well. Fracking's most immediate

"I'd give all the wealth that
years have piled, the slow
results of life's decay,
To be once more a little child
for one bright summer day."

—Lewis Carroll

threat to human health, however, is the massive amount of tainted wastewater it generates. By some estimates, fracking produces more than two hundred eighty billion gallons of wastewater annually. The environmental and health impacts of this drain on our precious water supply is still being tallied.

But fracking is only a stop-gap measure—the earth's supply of fossil fuels has been severely depleted.

Oil has been an essential tool in our lives (we drive our cars and we heat our homes with oil). But things must change because the sources are running low, and our fossil fuel dependence is damaging our earth.

Now we must invent a newer, better way to live. Our future must travel a different path than our past.

EXERCISE:
>   For one day, try to keep track of how much oil-derived energy you use.

AFFIRMATION:
>   *I acknowledge that fossil fuels are finite in supply.*

"The whole problem of industrial agriculture is putting all of your eggs in one basket. We need to diversify our food chains as well as our fields so that when some of them fail, we can still eat."

—MICHAEL POLLAN

6

THE FOOD THAT NOURISHES

"The word humility (also human) is derived from the Latin humus, meaning 'the soil.' Perhaps this is not simply because it entails stooping and returning to earthly origins, but also because, as we are rooted in this earth of everyday life, we find in it all the vitality and fertility unnoticed by people who merely tramp on across the surface, drawn by distant landscapes."

—PIERO FERRUCCI

FOOD NOURISHES OUR bodies and enriches our lives. Meals, cooking, and cherished recipes bring friends and families together. Food is meaningful to all of us, in many different ways.

No matter what we choose to eat, the food comes to us from our Earth. For this reason, we must respect the bounty of our planet's seasons. It is hard to imagine how our eating habits can have a negative effect on the Earth. But we are coming to learn that our insistence on a varied selection of food, year round, comes with a high environmental price tag.

Much of what we eat today comes to our tables from another climate. Today, walking through the aisles of a grocery store is almost like taking a trip around the world: strawberries from Chile and Asian pears are available in the middle of winter. Americans have come to expect constant variety in their diet, wherever they live, whatever the time of year. This means we must often import our food from

"He is your needs answered. He is your field which you sow with love and reap with thanksgiving. And He is your board and your fireside. For you come to him with your hunger, and you seek him for peace."

—KAHLIL GIBRAN

across the country or around the world. It is important to understand how this affects our Earth.

When food has to travel great distances, it is not just shipping that is expensive; the environmental impact is also costly. The true cost of transporting a flat of strawberries from Chile to a grocery store in the United States, for example, must also take into account the manpower, fossil fuel consumption, and carbon emissions required for those strawberries to make their journey from foreign fields to our table.

"Food miles" is the phrase used to designate the number of miles a product must travel in order to reach the consumer. A flat of strawberries from Chile shipped to Ohio has traveled more "food miles" than a flat from California. Experts argue that a measurement of foods impact on the Earth should also take into account *weather* (strawberries grown in hothouses have a bigger carbon footprint than those grown in sunlight) and *transportation type* (ocean freighters use less fuel than aircraft). "Food miles" are a rough, but crucial, calculation of the environmental impact of our

"The love of dirt is among the earliest of passions, as it is the latest . . . Fondness for the ground comes back to a man after he has run the round of pleasure and business. The love of digging in the ground (or of looking on while he pays another to dig) is sure to come back to him, as he is sure, at last, to go under the ground, and stay there."

—Charles Dudley Warner

foodstuffs that can help us assess the full costs of our food choices.

Another aspect of today's food production with negative environmental and health consequences is "factory farming" or "concentrated animal feeding operations" (CAFOs). Both terms refer to the increasing industrialization of the ways in which we raise and process our animal foodstuffs. From massive beef and pork operations with their tainted manure pools to overcrowded poultry farms, domestic food production in the United States is increasingly dominated by a handful of corporations. This has certainly resulted in lower food prices at the supermarket—Americans spend less than 10% of their disposable income on food. But it has also driven family farms out of business and polluted whole swathes of our countryside. Experts also contend that factory farming is a factor in the recent upswing in outbreaks of salmonella and other foodborne illnesses; when just a few operations are the source of our country's food supply, a single outbreak can have far-reaching consequences.

When we think in terms of food miles, those Chilean strawberries are expensive—even if

"The work of a garden bears visible fruits—in a world where most of our labors seem suspiciously meaningless."

—PAM BROWN

they cost less than fruit from California, the environmental cost is high. And that cheap beef or chicken comes with a hidden price tag of air and water pollution and unemployed family farmers. When our expectations for modern life lead to behavior that is wasteful, and ultimately unsustainable, we must make a change for the Earth.

EXERCISE:
> At your next meal, consider where each food item came from. Try to identify the sources.

AFFIRMATION:
> *The food I eat sustains and nourishes me.*

"The earth will not continue to offer its harvest, except with faithful stewardship. We cannot say we love the land and then take steps to destroy it for use by future generations."

—POPE JOHN PAUL II

7

THE FRAGILE
BALANCE

"Most of the observed increase in global average temperatures since the mid-twentieth century is very likely due to the observed increase in [human-caused] greenhouse gas concentrations."

—UNITED NATIONS INTERGOVERNMENTAL PANEL ON CLIMATE CHANGE (IPCC)

OUR PLANET CANNOT protect us if we do not protect our planet's atmosphere.

Beyond the clouds, our Earth is surrounded by "greenhouse gases," layers of elements like carbon dioxide, methane, and nitrous oxide. Together, these gases form a protective layer over our planet. It is important that these gases exist together in the right amounts. Otherwise, when that balance is disrupted, we experience global warming. This imbalance can raise average temperatures and cause rapid climate change.

What can cause the imbalance that leads to global warming? For one, the worldwide increase in carbon emissions from the burning of fossil fuels. The industrial boom occurring in China, India, and other developing countries worldwide is contributing to dangerous levels of pollution.

It's important to remember that this pollution is not restricted to the offending countries or regions. NASA has computer-generated models of planet Earth that illustrate this vividly—in real-time simulations, greenhouse gases swirl up from industrialized regions

"Earth, my dearest. Oh believe me, you no longer need your springtimes to win me over . . . Unspeakably, I have belonged to you, from the first."

—RAINER MARIA RILKE

and spin off across the globe, just the way the smoke from your neighbor's BBQ grill drifts into your yard. Global warming knows no national boundaries. China's pollution is our pollution, and vice versa.

Global warming means our Earth's delicate equilibrium is being thrown out of balance. Increased average temperatures have been blamed for shrinking polar ice caps, heat waves, droughts, and wildfires, as well as fierce tropical typhoons and hurricanes.

Experts agree that global warming is one of the most profound threats of our time. We must do everything in our power to guard the fragile balance of Earth's atmosphere.

EXERCISE:
> Walk outside and experience the climate. Is a breeze brushing your cheek? Is the sun warm on your skin? How does it feel to you?

AFFIRMATION:
> *I acknowledge that my choices today have a long term impact on the Earth's climate.*

A Prayer for the World

BY
AMY POWERS, ROB WELLS
AND CHARLENE GILLIAM

Can you imagine
Just for one moment in time
Every soul that's on this earth
Finds the silence
To go to that deep place inside
Where we know what life is worth

If we can only reconnect
The joy of what we have
Forget the things that we can't do
And just do what we can

"Action expresses priorities."

—MAHATMA GANDHI

7 SIMPLE
WAYS

"We need to rediscover the vast, harmonious, pattern of the natural world we are a part of—the infinite complexity and variety of its components, the miraculous simplicity of the whole."

—JAMES RAMSEY ULLMAN

`YOU

CAN SAVE THE EARTH

I n the course of a single day, every one of us makes hundreds of decisions. Many of these choices have an impact on our environment. Choices like: should I ride the bus or take my car? Should I shop at a local store or drive to the mall?

Make the effort to learn how to make better choices for our environment. Many of these choices are small and simple—yet powerful.

Change can be difficult, especially when life is so hectic. But when it comes to protecting our Earth for the future, change is essential. The choices we make today will shape the future of our Earth. The future

"Love the earth and sun and the animals . . . read these leaves in the open air every season of every year of your life."

—WALT WHITMAN

will be determined by the choices we make now, in our own homes, towns, and cities. Every time we make a decision, we have an opportunity to protect our environment. And although it may be difficult at first, it is only change that can alter the course of history.

Begin today.

". . . . when you work you fulfill a part of earth's furthest dream, assigned to you when that dream was born, And in keeping yourself with labour you are in truth loving life, And to love life through labor is to be intimate with life's inmost secret.

—KAHLIL GIBRAN

1

LOVE THE EARTH

"It really boils down to this: that all life is interrelated. We are all caught in an inescapable network of mutuality, tied into a single garment of destiny. Whatever affects one directly, affects all indirectly."

—Martin Luther King, Jr.

BEFORE WE CAN effectively change our actions, we must change our mindset. In everything we do, we must act out of love for the Earth. In the same way that you love a child, a friend, a family member, or anyone special in your life, you must also love the Earth. Your love requires thoughtfulness and action.

What does it mean to "love the Earth?" It means taking the time to think about what would protect our environment—and what might harm it—before we act. This is called "Earth-strategizing."

To Earth-strategize is to seek opportunities for personal practices that are Earth conscious. Here is an example. When you have a chore to do outside, like clearing the lawn of leaves, choose to rake instead of using your leaf blower. In this way, you are protecting the environment, and also using that time to appreciate the season's gifts. Our lives often move too fast for us to take the time to notice the wonder of the Earth. But we can make the time, and help our environment, too.

"The sun, the moon and the stars would have disappeared long ago . . . had they happened to be within the reach of predatory human hands."

—HAVELOCK ELLIS

Another part of Earth strategizing is practicing waste awareness. Simply put, this means being aware of the amount of waste we generate on a daily basis, and trying to cut it down. It's been estimated that the average person throws away approximately four pounds of garbage every day. By taking advantage of paperless billing and automatic bank deposits, we can take positive steps towards significantly reducing that amount. Waste awareness (like grabbing fewer paper napkins, skipping the ATM receipt, or bringing your own mug to work instead of using disposable cups) can sometimes feel like "baby steps". But cumulatively, the effect can be transformative. Once you start to practice waste awareness, you will look at things differently. The next time you open a package, you will be shocked by the amount of plastic, cardboard, and other materials that surround the contents. You will realize that after you open the package, all that material is useless, and you will suspect that such excessive packaging does not make good use of our resources or our planet. And you will be right.

The Environmental Protection Agency suggests that packaging (in this case,

"We abuse land because we regard it as a commodity belonging to us. When we see land as a community to which we belong, we may begin to use it with love and respect."

—ALDO LEOPOLD

food packaging) constitutes as much as one-third of the solid waste generated by individuals and households. What can you do to lower that statistic? Well, next time you need an item, look into purchasing it closer to home. Because ours is a small, blue planet with limited resources. We must protect it by restricting the amount of waste we generate.

Properly disposing of our waste is another component of waste awareness. It's safe to say that smart phones are now a regular part of our daily habits and lifestyle. But did you know that E-waste accounts for 2 to 5 percent of our solid waste stream? When improperly disposed of, E-waste (from computers, printers, and televisions to cell phones and electronic games) leaches harmful toxins like lead, mercury, and into our ecosystem. When disposing of E-waste, be sure to follow manufacturer guidelines and take advantage of municipal recycling events.

Earth-strategizing and making good use of the Earth can be tricky at first, but once we get used to it, it will seem obvious and natural. Whatever the activity, there is an Earth-friendly way to do it. If each one of us

"If you truly love nature, you'll find beauty everywhere."

—Vincent van Gogh

Earth-strategizes each and every day, we can move mountains.

We can become generous towards the Earth. We can learn to seek opportunities to make choices that are good for our Earth.

EXERCISE:
    See in your mind Earth as a living thing worthy of love.

AFFIRMATION:
    I love my life, I love my Earth.

"To see a world in a grain of sand,
And a heaven in a wild flower,
Hold infinity in the palm of your
hand,
And eternity in an hour."

—William Blake

2

MAKE WISER
CHOICES

"There are two ways to get enough: one is to continue to accumulate more and more. The other is to desire less."

—G.K. Chesterton

OUR LIVES ARE made up of many choices. With these choices come responsibility. We can decide to make better choices about what we buy and how we live. And we have a responsibility to do so. As a consumer, you are powerful: you decide what to buy and what companies to support. If you choose healthy and Earth-friendly products, manufacturers will supply them to us—and they will be cheaper—it is simply supply and demand.

Our best choices are made when we are informed. Make an effort to learn how your product got to the store. This means knowing who makes or grows the item, and where. Together, we can make organic, biodegradable, recycled, and Earth-friendly goods readily available and inexpensive. When it comes to food, this means becoming educated, responsible consumers.

Luckily, today's consumers have many tools at their disposable when it comes to determining the provenance and healthfulness of our foodstuffs. The "USDA Organic" label is one such tool. Agricultural products

"The secrets of this earth are not for all men to see, but only for those who will seek them."

—AYN RAND

bearing the "USDA Organic" label must meet stringent requirements concerning ingredients, production, and handling. Another useful tool is the Fair Trade Certified™ label. From herbs and spices to apparel and home goods, products certified as "Fair Trade" are grown, manufactured, and processed in socially and environmentally conscious ways, benefiting both producer and consumer. The Fair Trade system recognizes the mutual interdependence of our global economy—in the 21st century, we are all part of one great planetary market. And that market must be Earth-conscious in order for us all to thrive.

So the next time you shop, think carefully. Keep in mind the often-excessive food miles some of our imported food must travel to reach our stores. We have also reviewed the often unhealthy conditions under which our factory farms sometimes operate. That awareness should raise a red flag when you come across those imported strawberries or that discounted pork sausage, and you will look for strawberries or pork that are locally grown and produced, or choose another fruit or meat product that is local or in season.

"And if we do act, in however small a way, we don't have to wait for some grand utopian future. The future is an infinite succession of presents, and to live now as we think human beings should live, in defiance of all that is bad around us, is itself a marvelous victory."

—HOWARD ZINN

Keep learning. Research and understand the origins and practices that delivered that box of cereal or carton of eggs to the grocery store. Maybe the company that processed those eggs has been accused of unsanitary practices. Are you going to buy the eggs, no matter how inexpensive it is? No. Also try to understand the manufacturing conditions of the products. How environmentally sound is the plant? Does it pollute? The answers to these questions should influence your purchases. Be a responsible, conscientious consumer.

Finally, we also need to be concerned with where our products end up. Phosphates may biodegrade, but they can cause algae blooms when released into streams, rivers, and lakes. And just what does that "biodegradable" label mean on your household cleaner? It should mean that the cleaner contains organic ingredients that decompose easily.

Try the new green products. If they work for you, try some more. When it comes to saving the Earth, each individual initiative adds up, and is a step in the right direction. Pretty soon, it'll get harder and harder to find non-Earth-friendly products on the shelves.

"In the end, there is really nothing more important than taking care of the earth and letting it take care of you."

—CHARLES SCOTT

You won't be alone. According to a 2013 survey 71 percent of Americans consider the environment when they shop. As a bloc, Earth-aware consumers are increasingly dominating the marketplace. That will result in more and more green products on our store shelves. Again, it's simple supply and demand.

Remember, you wield immense power as a consumer.

EXERCISE:
Replace your cleaning products as well as soaps, detergents, and shampoos with Earth-friendly ones. If you don't understand the ingredients on a product label, don't buy it.

AFFIRMATION:
When I choose something Earth-friendly, I choose something good for me and my loved ones.

"If civilization has risen from the Stone Age, it can rise again from the Wastepaper Age."

—Jacques Barzun

3

CHOOSE TO
REDUCE

"Do not wait for leaders. Do it alone, person to person."

—MOTHER TERESA

WE MUST CHOOSE to reduce our use of fossil fuels.

Why? For one, fossil fuels like oil and coal are nonrenewable resources. *Nonrenewable* means we cannot replace what we've used; once its consumed, it's gone for good. Second, the burning of fossil fuels for energy damages our environment.

We will have to change our habits. Right now, our modern way of life is organized around the car. We know that automobile exhaust is 95% carbon dioxide, and that the typical passenger vehicle produces over 5 metric tons of carbon dioxide per year. The conclusion is clear: We need to reduce our car habit.

Carpooling is one solution. Sharing a ride with co-workers can be a fun opportunity to chat. Taking the bus is another way to get out and make your solitary commute a social occasion. Some people vow to take the bus on nice days to cut down on driving days—and to get some exercise to and from the bus stop as well. And the more people

"The earth we abuse and the living things we kill will, in the end, take their revenge; for in exploiting their presence we are diminishing our future."

—MARYA MANNES

taking the bus or train, the more the routes and schedules will reflect where and when we want to travel.

Bicycling is becoming a popular mode of transportation again. Many American cities have redesigned their streets and roads to accommodate dedicated bike lanes. Bike-sharing programs are also proliferating. Picking up a bike from one location and dropping it off in another is a great way to make short trips enjoyable and healthy. It's also a great way to economize—statistics suggest that the average consumer can save up to $800 a year in transportation costs by participating in a bike-sharing program.

For shorter distances, there is nothing better than walking. Medical experts agree that a brisk walk has innumerable benefits, including maintaining a healthy weight, preventing or managing conditions like heart disease and high blood pressure, and helping to strengthen bones and muscles. Not to mention that a good walk never fails to improve your mood! Consider shopping closer to home. Many of our small towns and cities have at least a shop or two that provide essentials and are just a walk or bike ride away. Often, we overlook

"Each and every master, regardless of the era or the place, heard the call and attained harmony with heaven and earth. There are many paths leading to the top of Mount Fuji, but there is only one summit—love."

—Morihei Ueshiba

these smaller stores for the bigger ones because larger chains offer more "deals." But how much money do we *really* save in the end by driving to a huge store? The trip itself will cost us gas money; not to mention the fact that a store with huge selection may tempt us into buying more than we need. Riding a bike to the store is also a great solution.

So next time, try walking or riding to a local store to pick up that carton of eggs. You will save yourself time and the headache of waiting in traffic and looking for a parking spot. You'll see the sights. And you'll get some exercise, too!

What we hope is that if we look a little closer at things, we just might find answers that will not only improve the Earth, but change our lives, too. By rethinking our priorities, we can rediscover a pace of life that is slower, more localized, more community-oriented. A way of life that is better for us, and better for the Earth.

"Study nature, love nature,
stay close to nature. It will
never fail you."

—FRANK LLOYD WRIGHT

EXERCISE:
Make a short list of ways you can reduce
your dependency on fossil fuel.

AFFIRMATION:
*I am committed to doing my part to con-
serve limited energy resources.*

"The system of nature, of which man is a part, tends to be self-balancing, self-adjusting, self-cleansing. Not so with technology."

—E.F. SCHUMACHER

4

EMBRACE GREEN TECHNOLOGY

"Humanity is on the march,
earth itself is left behind."

—DAVID EHRENFELD

OUR FUTURE WILL be built on "green" technology.

That future is here now.

More and more, manufacturers around the world are using technological innovation to make products work more on less energy.

Take this simple example: the light bulb. As you probably noticed on your last trip to the hardware store, the traditional incandescent light bulb has been phased out in the United States, as well as in Canada, Australia, and much of northern Europe. Incandescent lightbulbs have been replaced by energy-efficient alternatives, including LED (light-emitting diode) and compact fluorescent bulbs (CFLs) . . . and for good reason—LED light bulbs use one-quarter to one-third of the energy and last 8 to 25 times longer than halogen incandescents.

You might hear "fluorescent lighting" and automatically think of your office, the gym, or a doctor's office. But CFL lighting is very sophisticated, ranging from bright white to

"A margin of life is developed by Nature for all living things—including man. All life forms obey Nature's demands—except man, who has found ways of ignoring them."

—EUGENE M. POIROT

warm, like incandescent light. The bottom line is, by replacing our centuries-old incandescent bulbs, we can reduce our consumption of electricity and immediately reduce our carbon emissions. All that with only a twist of the wrist!

Every product you buy presents an opportunity for you to conserve energy and embrace green technology. Here again, today's Earth-friendly consumer has effective tools and information at her disposal, namely ENERGY STAR. An initiative of the Environmental Protection Agency (EPA), ENERGY STAR has been lauded as "the most successful voluntary energy conservation movement in history," helping to "identify and promote energy efficiency in products, homes and buildings nationwide." Energy-efficient appliances like microwaves, washers, dryers, and air conditioners can all be found with the ENERGY STAR logo—it's a dependable indicator of Earth-friendly products.

Heating and cooling our homes accounts for 48 percent of our monthly utility bills. So turn down the thermostat during the winter, and in hot weather, draw the blinds or curtains, open the windows on the shady side

"I realized that Eastern thought had somewhat more compassion for all living things. In the East, the wilderness has no evil connotation; it is thought of as an expression of the unity and harmony of the universe."

—William O. Douglas

of the house—and if you don't have shade, plant some trees. They will also protect your house from harsh winter winds. Every degree warmer or cooler represents big savings both for you and the planet.

There are other, more long-term ways to save energy as well. Consider installing solar panels or a solar water heater—every year, the price per watt of installing residential solar panels continues to drop, bringing this cutting-edge technology within reach of many consumers. Another way to save energy is reinsulating and resealing your home. By adding insulation in the attic, or in crawlspaces and replacing old-single paned windows with double-paned models, you can save substantially on your utility bills. You can also make changes to your hot-water heater. About 12 percent of an average home energy bill goes to heating water. By turning down the temperature of your hot water, insulating the tank, and replacing old heaters with more energy-efficient models, the average family can save energy and reduce their monthly fuel bills—and, in some states, get a tax rebate or other incentive. The average American household spent over $2,100 on

"Nature gives to every time and season some beauties of its own."

—CHARLES DICKENS

energy in 2012. Improvements to our homes' fuel efficiency can lower household fuel costs and significantly reduce the generation of greenhouse gas emissions.

Energy efficiency is the key to our future. Embrace it now.

EXERCISE:
Conduct a survey of your home and discover where you can apply green technology to conserve.

AFFIRMATION:
I will choose energy and resource-saving technology.

"We live in the world, and the world lives in us."

—ALBERT SCHWEITZER

5

RECYCLE, REUSE, REPAIR

"Sooner or later, we will have to recognize that the Earth has rights, too, to live without pollution. What mankind must know is that human beings cannot live without Mother Earth, but the planet can live without humans."

—EVO MORALES

WE HAVE A responsibility to the Earth to use our resources carefully.

One way to do this is by recycling. We can also get the most out of what we own by repairing appliances and products so they last, instead of buying new ones. We can also reuse items, like boxes, for other purposes.

By now, most of us are familiar with the importance of recycling. Hopefully, we do it without even thinking. But just because it has become second nature to most of us shouldn't blind us to the fact that recycling is making a major difference. If anything, the way recycling has become a part of our everyday lives should be an example of how to incorporate other environmentally-minded changes into our lifestyles.

Time has shown that recycling works. Since the 1970s, recycling programs have made use of literally tons of discarded material. Imagine: so-called garbage was successfully made useful again—instead of just taking up room in a dump. Nevertheless, we can and must step

"It appears to be a law that you cannot have a deep sympathy with both man and nature."

—HENRY DAVID THOREAU

up our efforts. The EPA estimates that 75% of the American waste stream is recyclable, but we only recycle about 30% of it.

Here's an incentive: did you know that the business of recycling is profitable, too? For one, it contributes to a healthy job market. It is estimated that sorting and recycling discarded materials generates 7 to 10 more jobs than landfills and waste to energy plants. The Office of the Federal Environmental Executive estimates that recycling and remanufacturing industries account for approximately 1 million manufacturing jobs and more than $100 billion in revenue. Companies are increasingly discovering that the cost of recycling is matched by profit earned from manufacturing with reused materials, rather than using new, raw materials exclusively. Increasingly, recycled materials are even becoming significantly cheaper to use than new materials.

Our individual recycling efforts can have major positive results. Here are two statistics that should encourage you. Over 87% of Americans have access to curbside or drop-off paper recycling programs. In the twenty years between 1990 and 2010, paper recycling

"Time and space—time to be alone, space to move about—these may well become the great scarcities of tomorrow."

—EDWIN WAY TEALE

increased over 89%. Both are clear indications that recycling is not only a possibility for communities nationally, but a top priority.

In addition to recycling, we can reuse and repair. This is easier than you might think. Imagine how much plastic you'll conserve, and how much money you'll save, by using refillable glass containers for water instead of buying water in plastic bottles. Or you can start using washable dishtowels instead of paper towels. Be less wasteful.

We just have to change the way we think. Historically, people raised during tough times knew how to make due. Many individuals who were raised during the Great Depression pride themselves on their ability to fix *anything*, and reuse common items that the rest of us would throw away, such as using the bags that baked bread comes in for plastic baggies. The wasteful habits we are trying to change are recent ones, and we can overcome them by remembering and honoring our parents' and grandparents' resourceful ways of living.

Recycling, reusing and repairing are part of a simple shift in our behavior that will change our impact on the Earth.

"The miracle is not to walk on water. The miracle is to walk on the green earth in the present moment, to appreciate the peace and beauty that are available now."

—THICH NHAT HANH

EXERCISE:

Consider all the material possessions you own. Choose to take as much care as needed to extend the life of these items. Within the next month, sew a tear, pass on unneeded items to others, and find ways to encourage recycling at home or at work.

AFFIRMATION:

I will be mindful of the goods I buy and keep.

"What is the use of a house if you haven't got a tolerable planet to put it on?"

—HENRY DAVID THOREAU

6

THINK LOCAL

"The land and sea, the animals, fishes, and birds, the sky of heaven and the orbs, the forests, mountains, and rivers, are not small themes."

—WALT WHITMAN

WHAT DOES IT mean to think local? "Local" means neighbors. "Local" means you can walk or ride a bike or bus to wherever you need to go. "Local" means less energy consumed, and therefore fewer carbon emissions. "Local" means being kinder to our planet.

As large as our world has grown, it's often still possible to find everything we need within miles of our homes. All we have to do is look.

Why is "local" better for the planet? When you buy locally, you save energy. But just as importantly, you promote regional farmers and businesses . . . which means you're helping to build a stronger community. How can you buy local? Farmers markets and food stands are springing up all across the country, and are a perfect opportunity to find some really great produce and meet the man or woman who grew the tomato or gathered the egg you're buying. Knowing the people who help put food on your table also builds a stronger community.

Buying locally also means that the food you

"We shall never achieve harmony with land, any more than we shall achieve absolute justice or liberty for people. In these higher aspirations, the important thing is not to achieve but to strive."

—ALDO LEOPOLD

serve your family will be fresh, in season, and unique to the region you live in. Many cooks will tell you that it is wonderful to have access to imported tomatoes in the middle of winter. But they will also admit those imported tomatoes do not taste anything like the summer tomatoes from your garden or your local farmers market. Buying food locally usually means quality you can see and taste.

There's a saying that all politics is local. We think this means that no matter what they say or do in our state capitols or Washington, it's how we deal with issues in our own backyards that will make a change. "Local" means community. So support your local community organizations. When we think, buy, live, and act locally, we strengthen our community— and it is as a community, working together, that we will save the Earth.

"When we try to pick out anything by itself, we find it hitched to everything else in the Universe."

—JOHN MUIR

EXERCISE:
Take one week to discover local sources for your daily needs: food, entertainment, services, and other necessities and tasks.

AFFIRMATION:
I choose to be aware of the myriad opportunities to meet my needs within my community.

"What would the world
 be, once bereft
 Of wet and of wildness?
 Let them be left,
 O let them be left, wildness
 and wet;
 Long live the weeds and the
 wilderness yet."

—GERARD MANLEY HOPKINS

7

CHERISH THE EARTH'S PRECIOUS GIFTS

"We never know the worth of water till the well is dry."

—THOMAS FULLER

EACH OF US must play our own part in helping to save the Earth.

As you work toward making a difference, remember why you want to save our planet: perhaps it's the simple joy you feel in the taste of fresh water, or in the touch of a sweet smelling breeze. Or maybe it's the sense of solidarity you feel with the Earth when you look at an awe-inspiring landscape unspoiled by human development.

So how can you cherish the Earth?

Look outside your home. Whether you live in the country, the suburbs, or the city, seek air, trees, grass, and sun. Embrace the explorer within by taking a road you've never been down just to see where it leads. Go for a hike with your kids or a friend, or ride your bike to a neighboring town. Plan a vacation to the seashore to learn about creatures great and small, from whales to the tiny life forms that dwell in tidal pools. On the weekend, wake up early on Saturday and go fishing. You may not catch anything, but you'll be

"Man has been endowed with reason, with the power to create, so that he can add to what he's been given. But up to now he hasn't been a creator, only a destroyer. Forests keep disappearing, rivers dry up, wild life's become extinct, the climate's ruined and the land grows poorer and uglier every day."

—ANTON CHEKHOV

able to watch mist rise off the fields as the sun comes up.

Look in your own backyard, too. You can plant a vegetable garden (using your own compost!) or even fill planters with flowers. If you live in a big city, join a community garden. Even home improvements can be meaningful. Build a tire swing for your grandkids or contribute one to your local park, so kids can swing and pretend to touch the sky.

We know more about our Earth and its wonders than we ever have, and we are discovering more all the time. All our knowledge should only make us more protective, and more awestruck, at the beauty of our home. This great planet has most value in the human eye and heart. So explore the woods. Lay in the grass and watch the clouds gather overhead. Hike to the top of a hill and breathe deeply.

No matter what form your connection to our planet may take, the importance of the relationship between humans and the Earth is undeniable, and must never be forgotten.

We will save the Earth by loving it.

"To find the universal elements enough;
to find the air and the water
exhilarating; to be refreshed
by a morning walk or an
evening saunter; to be thrilled
by the stars at night;
to be elated over a bird's nest
or a wildflower in spring—
these are some of the rewards
of the simple life."

—JOHN BURROUGHS

EXERCISE:
   The next time you take a walk, take a
   hike, go for a bike ride, or go for a swim,
   bring a friend or family member along.
   Enjoy the day. Talk about what you see in
   nature around you. Smile. You are alive!

AFFIRMATION:
   *Feel gratitude for the Earth's resources
   in everything you do. Say, "thank you."*

A SPECIAL THANK YOU

You've reached the last page of this little book—now it is up to you to write the next chapter. Take the lessons you have learned here and share your discoveries with others. Be a force for sustainability in your home, community, and workplace. Choose to lend a hand, plant a tree, and make a difference.

ANDREW FLACH, Publisher
RYAN TUMAMBING, Associate Publisher
ANNA KRUSINSKI, Managing Editor
RYAN KENNEDY, Assistant Editor
SEAN SMITH, Writer

NOTES

PREFACE

"Lyndon B. Johnson: Remarks at the Signing of a Bill Establishing the Assateague Island Seashore National Park." *Lyndon B. Johnson: Remarks at the Signing of a Bill Establishing the Assateague Island Seashore National Park.* N.p., n.d. Web. 23 Sept. 2016.

INTRODUCTION TO PART 1

"Remarks by the President at U.N. Climate Change Summit." *The White House.* The White House, 2014. Web. 23 Sept. 2016.

Aldo Leopold, "Conservation" (c. 1938); Published in *Round River,* Luna B. Leopold (ed.), Oxford University Press, 1966.

Wendell Berry, *The Unsettling of America: Culture & Agriculture.* San Francisco: Sierra Club Books, 1977.

Edward Abbey, *Down the River.* New York: Dutton, 1982.

William Butler Yeats, *"The Municipal Gallery Revisited," The Collected Poems of W.B. Yeats.* London: Wordsworth Poetry Library, 2000.

Walt Bresette, "PTE." http://www.protecttheearth.org/Walter/aboutwalt7.htm N.p., n.d. Web. 23 Sept. 2016.

CHAPTER 1

John Muir, *My First Summer in the Sierra.* Boston: Houghton Mifflin, 1911.

"Indonesian Fires Girdled Half the Globe in Smoke." *National Geographic.* National Geographic Society, n.d. Web. 23 Sept. 2016.

Annie Dillard, *The Writing Life* (New York: Harper & Row, 1989).

Tom Philpott. "The Gulf of Mexico Is about to Experience a 'Dead Zone' the Size of Connecticut." *Mother Jones*. N.p., n.d. Web. 24 Sept. 2016.

"The Effects: Environment." *EPA*. Environmental Protection Agency, n.d. Web. 24 Sept. 2016.

Al Gore, *An Inconvenient Truth*. Film. (2006)

Ralph Waldo Emerson, Conduct of Life: "Power" (1860). http://www.emersoncentral.com/power.htm

CHAPTER 2

"WHO | Climate Change and Human Health - Risks and Responses. Summary." WHO | Climate Change and Human Health - Risks and Responses. Summary. World Health Organization, n.d. Web. 24 Sept. 2016.

Ibid.

"Jane's Prayer for Pope's New Declared Day for the Environment - Jane Goodall's Good for All News." Jane Goodall's Good for All News. http://news.janegoodall. org/2015/09/01/janes-prayer-for-popes-new-declared-day-for-the-environment/ N.p., 2015. Web. 24 Sept. 2016.

"A Ruined Planet Cannot Sustain Human Lives in Good Health." World Health Organization. World Health Organization, n.d. Web. 24 Sept. 2016.

CHAPTER 3

http://blog.americanrivers.org/wordpress/?s=cousteau; Lindsay Martin, Web Editor

"Bottled Water." http://www.bottledwater.org/economics/ bottled-water-market N.p., n.d. Web. 24 Sept. 2016.

Kahlil Gibran. *The Complete Works of Kahlil Gibran: All Poems and Short Stories*. New Delhi: General Press, 2016.

"Bottled Water Facts." Ban the Bottle RSS. N.p., n.d. Web. 24 Sept. 2016.

"Why Is Our Water in Trouble?" Rivers & Lakes Impacts. http://www.nature.org/ourinitiatives/habitats/riverslakes/ threatsimpacts/ N.p., n.d. Web. 24 Sept. 2016.

"What is Ocean Acidification?" PMEL Carbon Program. http://www.pmel.noaa.gov/co2/story/ What+is+Ocean+Acidification%3F N.p., n.d. Web. 24 Sept. 2016.

Hal Borland. *Twelve Moons of the Year*. New York: Alfred A. Knopf, 1979.

Al Gore, Nobel Lecture, 10 December 2007.

https://www.nobelprize.org/nobel_prizes/peace/ laureates/2007/gore-lecture_en.html

John Muir, *John of the Mountains: The Unpublished Journals of John Muir*, (1938), edited by Linnie Marsh Wolfe, (Madison: University of Wisconsin Press, 1938, republished 1979, page 439.

CHAPTER 4

Chief Seattle speech (1854), transcribed and published by Henry A. Smith, *Seattle Sunday Star*, October 29, 1887.

"Global Greenhouse Gas Emissions Data." EPA. Environmental Protection Agency, n.d. Web. 24 Sept. 2016. https://www.epa.gov/ghgemissions/ global-greenhouse-gas-emissions-data#Country

"The National Academies." Our Energy Sources. Fossil Fuels. http://needtoknow.nas.edu/energy/energy-sources/ fossil-fuels/ N.p., n.d. Web. 24 Sept. 2016.

"The National Academies." How We Use Energy, Transportation http://needtoknow.nas.edu/energy/ener-gy-use/transportation/. N.p., n.d. Web. 24 Sept. 2016.

U.S. Environmental Protection Agency, "Emission Facts: Greenhouse Gas Emissions from a Typical Passenger Vehicle" https://www.epa.gov/sites/production/files/2016-02/ documents/420f14040a.pdf

William Shakespeare, *Hamlet*, 2.2

"Overview of Greenhouse Gases." EPA. Environmental Protection Agency, n.d. Web. 24 Sept. 2016. https://www.epa.gov/ghgemissions/overview-greenhouse-gases

Wendell Berry. *The Long-Legged House*. Berkeley: Counterpoint, 2012.

Amory B. Lovins. "Ending Our Oil Dependence," *The Ripon Forum*, Volume 39, Number II (March/April 2005), p. 12.

CHAPTER 5

Dictionary of Energy, edited by Cutler J. Cleveland, Christopher G. Morris. Oxford: Elsevier, 2009.

Richard Heinberg, MuseLetter, no.185, September 2007. http://www.richardheinberg.com/museletter/185; http://old.globalpublicmedia.com/richard_heinbergs_museletter_peak_everything

Blaise Pascal, *Thoughts*. Edited by Charles Eliot. New York: P.F. Collier and Son, 1910.

Merriam-Webster. http://www.merriam-webster.com/dictionary/fracking n.d. Web. 25 Sept. 2016

"Fracking Pollutes Some Water, But Harm Is Not Widespread, EPA Says." *National Geographic*. National Geographic Society, n.d. Web. 25 Sept. 2016. http://news.nationalgeographic.com/2015/06/150604-fracking-EPA-water-wells-oil-gas-hydrology-poison-toxic-drinking/

Environment America: *Fracking by the Numbers*. http://www.environmentamerica.org/sites/environment/files/reports/EA_FrackingNumbers_scrn.pdf

Michael Pollan, "Michael Pollan Fixes Dinner." Mother Jones. Web. 25 Sept. 2016. http://www.motherjones.com/media/2009/02/michael-pollan-fixes-dinner

CHAPTER 6

Piero Ferrucci. *Inevitable grace: breakthroughs in the lives of great men and women: guides to your self-realization.* (Los Angeles: J. P. Tarcher, 1990)

Kahlil Gibran. *The Prophet.* (New York: Knopf, 1952).

Charles Dudley Warner. *My summer in the garden.* (London: Sampson, Low, Marston & Co., 1883).

USDA ERS - Chart: On Average, Americans Spend Just under 10 Percent of Their Incomes on Food. http://www. ers.usda.gov/data-products/chart-gallery/detail.aspx?chartId=52216 N.p., n.d. Web. 25 Sept. 2016.

CHAPTER 7

IPCC, 2007: Summary for Policymakers. In: Climate Change 2007: The Physical Science Basis. Contribution of Working Group I to the Fourth Assessment Report of the Intergovernmental Panel on Climate Change [Solomon, S., D. Qin, M. Manning, Z. Chen, M. Marquis, K.B. Averyt, M.Tignor and H.L. Miller (eds.)]. (United Kingdom and New York: Cambridge University Press, 2007).

Al Gore. *An Inconvenient Truth: The Planetary Emergency of Global Warming and What We Can Do About It.* (New York: Rodale, 2006).

NASA. https://www.nasa.gov/press/goddard/2014/november/nasa-computer-model-provides-a-new-portrait-of-carbon-dioxide/#.V-gxCvkrLIU Web. 25 Sept. 2016

James Ramsey Ullman. *The Age of Mountaineering.* (Philadelphia: Lippincott, 1954).

Mohandas Mahatma Gandhi. http://thinkexist.com/quotation/action_expresses_

priorities/215751.html

7 SIMPLE WAYS

Rainer Maria Rilke. *Duino Elegies & The Sonnets to Orpheus*. (New York: Random House, 2009).

CHAPTER 1

"Remaining Awake Through a Great Revolution," (Speech, March 31, 1968) in Clayborne Carson and Peter Holloran, eds. *A Knock at Midnight: Inspiration from the Great Sermons of Reverend Martin Luther King, Jr.* (New York: Warner Books, 1998).

"Center for Sustainability & Commerce." How Much Do We Waste Daily? https://center.sustainability.duke.edu/ resources/green-facts-consumers/how-much-do-we-waste-daily n.d. Web. 25 Sept. 2016.

"Food Packaging and Its Environmental Impact." - IFT.org. http://www.ift.org/knowledge-center/read-ift-publications/ science-reports/scientific-status-summaries/editorial/ food-packaging-and-its-environmental-impact.aspx. N.p., n.d. Web. 25 Sept. 2016.

"E-Waste Recycling Facts and Figures." *The Balance*. https://www.thebalance.com/e-waste-recycling-facts-and-figures-2878189. N.p., n.d. Web. 25 Sept. 2016.

William Blake. "Auguries of Innocence." In *The Pickering Manuscript* (c. 1803). http://www.blakearchive.org/exist/ blake/archive/work.xq?workid=bb126&java=no

CHAPTER 2

G.K. Chesterton, "On Running After Ones Hat," *All Things Considered* (1908).

"Environmental Leader." RSS. http://www. environmentalleader.com/2013/04/03/71-of-consumers-think-green-when-purchasing/. Web. 25 Sept. 2016.

CHAPTER 3

Mother Teresa http://thinkexist.com/quotation/do_not_wait_for_leaders- do_it_alone-person_to/214997.html

Eco. Coach. "How bicycle sharing systems can positively impact your company." Posted on August 7, 2013 by Lee Hager http://www.eco-coach.com/blog/2013/08/07/how-bicycle-sharing-systems-can-positively-impact-your-company/ Web. 25 September 2016.

CHAPTER 4

David W. Ehrenfield. *The Arrogance of Humanism*. (London: Oxford University Press, 1981)

Energy.gov. "Lighting Choices to Save You Money." http://energy.gov/energysaver/lighting-choices-save-you-money. Web. 25 September 2016.

The ENERGY STAR Story. https://www.energystar.gov/about

Energy.gov. "Heating and Cooling." http://energy.gov/public-services/homes/heating-cooling Web. 25 September 2016.

Energy.gov. "15 Ways to Save On Your Water Heating Bill." Allison Casey, 2009. http://energy.gov/energysaver/articles/15-ways-save-your-water-heating-bill Web. 25 September 2016.

Alliance to Save Energy. *Top 10 Home Energy Efficiency Tips For Earth Day*

Release Date: Wednesday, April 18, 2012 https://www.ase.org/news/top-10-home-energy-efficiency-tips-earth-day Web. 25 September 2016.

Albert Schweitzer, *Thoughts for Our Times*. Edited by Erica Anderson. The Albert Schweitzer Fellowship. 1975.

CHAPTER 5

John Vidal. "Bolivia's Defiant Leader Sets Radical Tone at Cancún Climate Talks." *The Guardian*. 2010.

https://www.theguardian.com/environment/2010/dec/11/cancun-talks-evo-morales Web. 25 Sept. 2016.

DoSomething.org. *11 Facts about Recycling*. https://www.dosomething.org/us/facts/11-facts-about-recycling Web. 25 Sept. 2016.

Recycle Across America. *"Get The Facts."* http://www.recycleacrossamerica.org/recycling-facts Web. 25 Sept. 2016.

Institute for Local Self-Reliance." *Recycling Means Business*." https://ilsr.org/recycling-means-business/ (2002) Web. 25 Sept. 2016.

U.S. Environmental Protection Agency, "Jobs Through Recycling: Economic Benefits." http://www.epa.gov/jtr/ econ/index.htm

DoSomething.org. *11 Facts about Recycling*. https://www.dosomething.org/us/facts/11-facts-about-recycling Web. 25 Sept. 2016.

Ibid.

CHAPTER 6

Henry David Thoreau. The Writings of Henry David Thoreau, Volume 11. (Cambridge: Riverside Press, 1894).

CHAPTER 7

Gerard Manley Hopkins, "Inversnaid," Poems of Gerard Manley Hopkins (London: Humphrey Milford, 1918). Also New York: Bartleby.Com, 1999. http://www. bartleby.com/122/33.html